USING AI

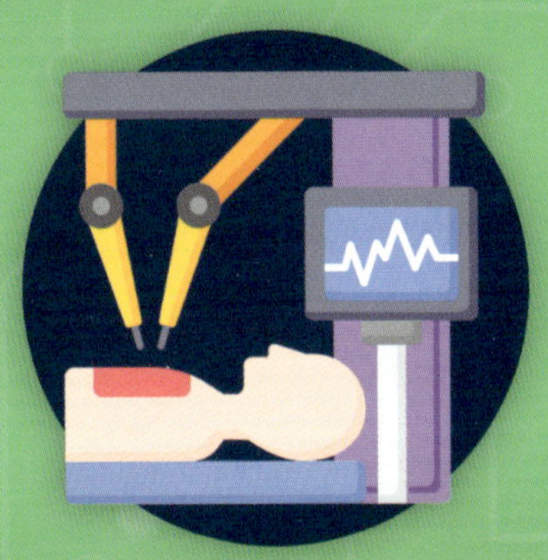

NANCY DICKMANN

BROWN BEAR BOOKS

Published by Brown Bear Books Ltd
4877 N. Circulo Bujia, Tucson, AZ 85718
USA
and
Studio G14, Regent Studios, 1 Thane Villas,
London N7 7PH, UK

ISBN 978-1-83572-046-2 (library binding)
ISBN 978-1-83572-052-3 (paperback)
ISBN 978-1-83572-058-5 (ebook)

Library of Congress Cataloging-in-Publication Data available on request

Text: Nancy Dickmann
Consultant: Matthew Lugg
Design and Illustrations: Square and Circus
Design Manager: Keith Davis
Children's Publisher: Anne O'Daly

Manufactured in the United States of America
CPSIA compliance information: Batch#AG/5666

Picture Credits
The photographs in this book are used by permission and through the courtesy of:
Cover: Freepik.com: **Interior:** iStock: Guillem de Balanzo 20–21, RGStudio 18–19; NASA: JPL-Caltech 12–13; Shutterstock: IM Imagery 14–15, Monkey Business Images 6–7, SeventyFour 8–9, Tong Stocker 10–11, Wavebreak Media 4–5, Drazen Zigic 16–17.
Artwork: Freepik.com.
All other artwork and photography
© Brown Bear Books.

Brown Bear Books has made every attempt to contact the copyright holder. If you have any information about omissions please contact: licensing@brownbearbooks.co.uk.

Websites
The website addresses in this book were valid at the time of going to press. However, it is possible that contents or addresses may change following publication of this book. No responsibility for any such changes can be accepted by the author or the publisher. Readers should be supervised when they access the Internet.

Words in **bold** appear in the Glossary on page 23.

Manufactured in the United States of America
CPSIA compliance information: Batch#AG/5666

CONTENTS

WHAT IS AI?

Our world is changing all the time. People invent new products. They find better ways of doing things. Long ago, we had no computers. Now many people use them every day. We use laptops, tablets, and smartphones. They make many jobs easier.

We use computers at home and at school. Many adults use them at work.

COMPUTERS THAT THINK

Computers today can be really powerful. Scientists are working on ways to make them think and learn. They want computers to be more like human brains. This is artificial intelligence. We call it AI for short. AI is still new. But it's becoming part of our daily lives.

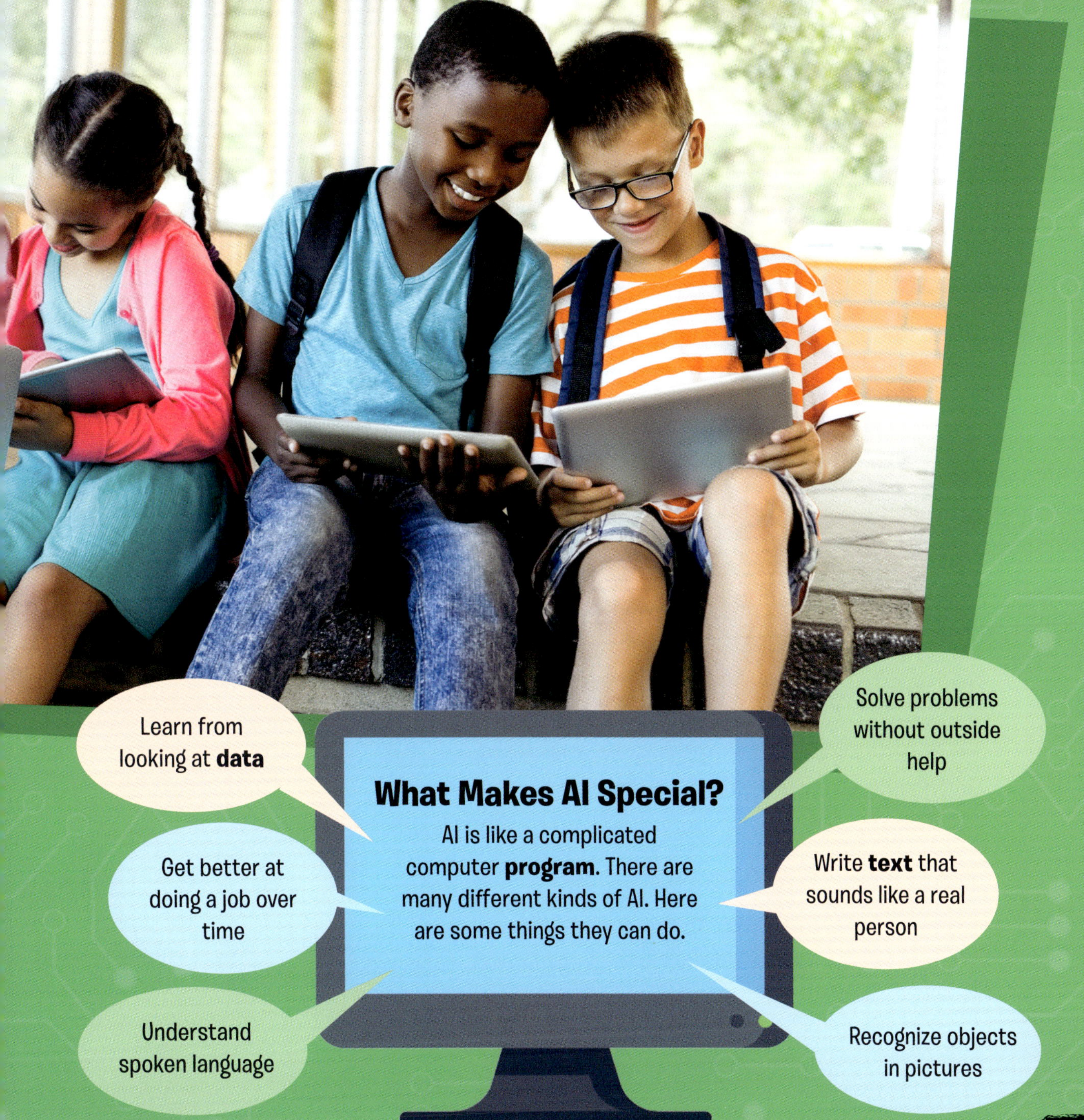

What Makes AI Special?

AI is like a complicated computer **program**. There are many different kinds of AI. Here are some things they can do.

- Learn from looking at **data**
- Get better at doing a job over time
- Understand spoken language
- Solve problems without outside help
- Write **text** that sounds like a real person
- Recognize objects in pictures

WHY WE USE AI

Call centers help solve problems for customers. AI might take jobs like these.

When AI is used well, it can be a big help. AI is very good at some jobs. It can even do them better or faster than a human could. Using AI can help companies save money. It can help them offer better products and services.

CRUNCHING NUMBERS

AI is really good at analyzing lots of data quickly. It teaches itself to find **patterns** in it. Humans might take longer to see these patterns. They could even miss them entirely. But AI can find these patterns fast. Then we can use what they've learned.

AI That Helps

AI tools can help people with disabilities. People who have trouble moving can use voice commands. The commands can turn on lights or send messages. AI can also read text out loud. This helps people who are blind.

What AI Can Do

Not all jobs are right for AI. Some need people.

Humans are better at:

- being creative, like artists and musicians
- working with their hands, like plumbers
- getting the best out of people, like coaches

AI can help with:

- analyzing tables and numbers
- filling in forms and sorting emails
- recognizing faces for security

HEALTHCARE

Doctors use AI to help their patients. AI can check **scans**. It looks for broken bones in x-rays. It checks for signs of disease. Sometimes it sees things that doctors missed. AI doesn't always get things right. But it learns from its mistakes.

FINDING NEW MEDICINES

Scientists search for new medicines. These medicines might kill germs or fight cancer. AI can help. It can **predict** what substances will to do germs. It makes computer models showing what might happen. This gives doctors a shorter list of substances to try. It speeds up the process.

Each patient is different. AI can help doctors choose the right treatment.

AI in Hospitals

AI does more than look at scans. It can help hospitals run better. Here are some examples.

AI programs order the right supplies at the right time.

AI-powered **robots** help with surgeries.

Chatbots help make appointments and give advice.

AI can predict which diseases might spread. Doctors can protect against them.

FARMING

On a farm, people grow crops. They raise animals. This might not sound like a place for AI. But farmers are starting to use it. Some farms don't have enough workers. AI tools can help share the load. AI can even drive a tractor!

GROWING CROPS

AI tools help farmers grow better crops. They get the best from their land. AI can track weather. It tells the farmer when to water or add plant food. AI tools can tell when crops are ready to pick. They can predict which crops will grow well.

Drones help on some farms. They check crops or spray chemicals.

Spotting Pests

Some AI can recognize bugs or weeds. Farmers take a photo of what they find. The AI tells them what it is. It says whether it is harmful. Then it suggests what to do about it.

AI on the Farm

Farmers use AI in many different ways. It helps them grow better quality food. They produce more of it.

AI spots weeds. It tells farmers exactly where to put weedkiller.

AI lets machines drive themselves. They can plant or pick crops.

Drones check the health of crops. They look for pests.

Some vegetables aren't good quality. AI-powered machines can sort them.

AI programs tell the farmer what the crops need.

EXPLORING SPACE

This rover studies rocks on Mars. AI tools help it decide which ones need a closer look.

Using AI helps us learn more about space. Scientists use AI to plan routes for spacecraft. AI is good at testing flight paths. It maps where there is space junk in the way. AI helps figure out how much fuel is needed.

TOO MUCH DATA?

Spacecraft send back loads of information. So do **telescopes**. Going through all this data takes a long time. AI can speed up the process. It can look for patterns. These might show new objects in the sky. They can show distant stars dying. AI spots things that humans might miss.

Finding New Planets

Some distant stars have planets around them. These are called exoplanets. AI can help find them.

Telescopes take pictures of the same stars. They take them at different times.

Scientists feed the pictures to an AI.

The AI checks the same star in each one.

Is the brightness always the same? It probably has no planets.

Does it sometimes get dimmer? There might be a planet passing in front of it and blocking the light.

FACTORIES

We buy things all the time. Most are made in factories. Some factories use robots to do the work. And some use AI! AI can help make factories run better. It keeps track of schedules. It predicts what orders will come in. Then it makes sure the right materials are available.

SEEING PROBLEMS

Factory machines sometimes break down. This can cause delays. AI can predict when a machine will need repair. Then workers fix it before it breaks. The AI gets data from the machine. It sees which parts are getting worn out. This helps it know when parts need replacing.

Robots are good at putting parts together. AI helps them make their own decisions.

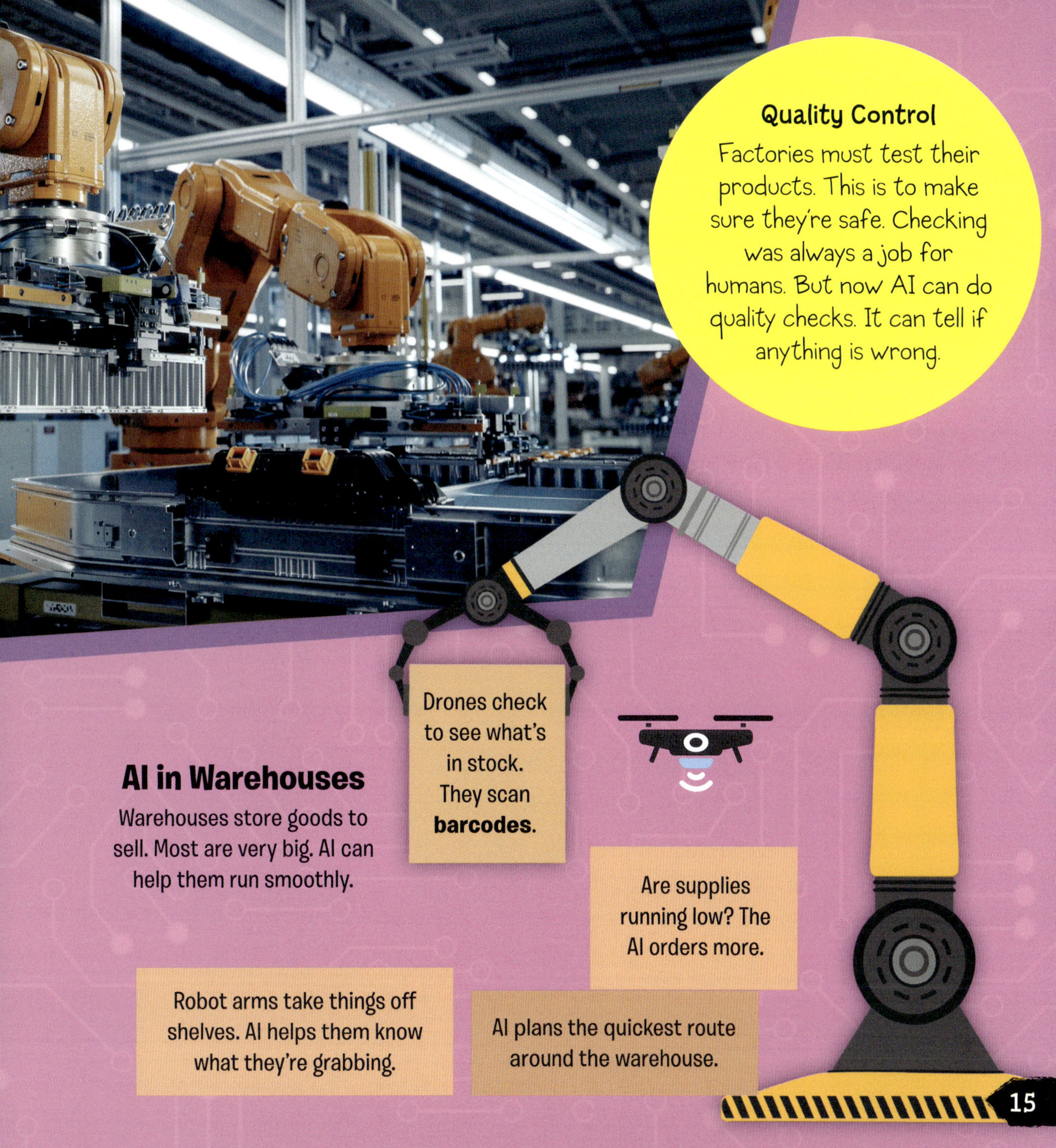

Quality Control

Factories must test their products. This is to make sure they're safe. Checking was always a job for humans. But now AI can do quality checks. It can tell if anything is wrong.

AI in Warehouses

Warehouses store goods to sell. Most are very big. AI can help them run smoothly.

Drones check to see what's in stock. They scan **barcodes**.

Are supplies running low? The AI orders more.

Robot arms take things off shelves. AI helps them know what they're grabbing.

AI plans the quickest route around the warehouse.

BANKING

Banks help us manage our money. AI chatbots can help them do this. They "talk" to customers. They can give information about the customer's account. They can help them solve problems. Some problems are too much for the AI to solve. Then a real person steps in.

LENDING MONEY

Banks often lend people money. They want to make sure the person will be able to pay it back. AI helps a bank do checks on borrowers. It looks at income and spending. It checks if they have other loans. This helps it predict if a person is safe to lend to.

Bankers work to build trusting relationships with their customers. AI cannot do this.

Catching Thieves

Bank cards might get stolen. They could be copied. AI helps banks spot when this happens.

The AI has records of how you use your card.

It looks for patterns. What do you buy? Where do you buy it?

Someone buys something. It doesn't match the pattern. It might be too expensive. It might be in a different country.

The AI tells the bank. The bank checks with the customer.

SHOPPING

We shop for food, clothes, and other things. AI helps stores meet our needs. Some stores have smart shelves. They know how much is on them. They can tell what needs re-ordering. Stores also use AI to predict what people will buy. Then the stores make sure they have it.

ONLINE SHOPPING

Do you ever shop online? You search for what you want. The website might also show you other things. This is often powered by AI. It tracks what you search for. It sees what you buy. Then it suggests similar products. You are more likely to want them.

Image Searching

If you see something cool, you can take a picture of it. Then you upload it. An AI finds a match on a shopping site. It finds similar products too. Then you can buy your own.

The Right Price

AI often helps set the prices in online shops. This is what it looks at.

ENTERTAINMENT

With streaming, we can watch whatever we want. But there are a lot of films and shows to choose from! Streaming services help by making suggestions. These are powered by AI. They track what you watch. They learn what you like. Then they suggest similar shows.

RATINGS

After you watch something, you can give it a rating. Did you like it or not? These ratings help the AI learn more about what you like. Say you didn't like a film it suggested. It will try to figure out why. In the future, its predictions will be better.

Music streaming uses AI to look for patterns in the music you like. It suggests other artists to try.

Making Entertainment

AI is also used to create **content**. Here are some examples of what it can do.

Create music tracks in different styles

Make special effects for films and TV

Help write scripts for shows and video games

Create animations for videos

Provide subtitles in different languages

Turn books into audiobooks

QUIZ

How much have you learned about the ways we use AI? It's time to test your knowledge!

1. Which of these is a way that AI helps doctors?

a. handing out pills to patients
b. looking at scans to spot problems
c. checking temperatures

2. Which of these can an AI drone not do?

a. check crops for pests and disease
b. spray chemicals to help crops
c. pull up weeds

3. Why is AI good at finding new planets?

a. it can go through lots of data quickly
b. it sees better than telescopes do
c. it designs its own spacecraft

4. Which of these purchases tells an AI that a thief might have stolen your card?

a. something you bought before
b. something in a different country
c. something that a friend recommended

The answers are on page 24.

GLOSSARY

barcode a pattern of lines and spaces printed on a product or on its packaging

chatbot a computer program designed to have conversations with humans

content material meant to be consumed online, such as text or videos

data information that is stored or used in a computer, in the form of a series of ones and zeroes

drone a flying machine guided by remote control or an onboard computer

pattern an arrangement of similarities or trends between different items in a set

predict to make an educated guess about what will happen

program a set of coded instructions for a computer to follow

robot a machine with moving parts that is programmed to do a job

scan an image taken of part of the body to help doctors spot injury or disease

telescope a scientific tool for looking at things that are far away, such as in space

text data in the form of written words

FIND OUT MORE

Books

AI Basics. Elsie Olson, Lerner Publications, 2025.

Artificial Intelligence. Julie Murray, Abdo Books, 2021.

How AI Works. Lisa Idzikowski, Lerner Publications, 2025.

Websites

www.bbc.co.uk/newsround/49274918

kids.britannica.com/kids/article/artificial-intelligence/390648

www.softwareacademy.co.uk/ai-for-kids/

INDEX

Answers: 1. b; 2. c; 3. a; 4. b